How Artificial Intelligence and Chatbots Have Changed Human–Robot Interaction

Tania Peitzker

Abstract: This article attempts to show by example the various forms of chatbots that have evolved since the 1950s when the British computer scientist, Alan Turing, first theorized about the "imitation game." He thought it was possible to recreate a human-like intelligence within a machine, so that it not only computed highly complex tasks impossible for a human brain to achieve, but also built a specific relationship with a person by simulating feelings, emotional intelligence, and overall personality. So, it was conceptualized as an evolving form of artificial intelligence refined by machines to better interact with humans. I explore how that initial aim of the now famous twentieth-century "Turing Test" when "a layperson chatting with a machine/chatbot for 5 minutes would be confused as to the human or robot nature of the source emanating from the computerised interface"—as first conceived by Turing's predecessors a century earlier with Ada Lovelace and Charles Babbage creating machines for this purpose in the early 1800s—has become the Holy Grail of Internet corporations and smart device manufacturers in our Internet of Things or *Industrie 4.0* times.

Given the difficulties of taking natural language processing chatbots to market, especially their speech recognition (ASR) counterparts such as Virtual Assistants, it has been necessary to look at a variety of Use Cases and industry trials on a global scale. I am aware of the great confusion amongst consumers, academics, and even chatbot developers and robot builders themselves as to the jargon and far-reaching concepts like AI. I address the need to refer to this emerging tech with consistency in terminology; thus the first part of the White Paper goes into the forms and classifications of AI, (chat)bots and the diverse algorithms that power them, often overlapping as hybrid source codes and software, even plug ins like auto-translation apps, which can vastly improve human–robot interaction.

In the past 6 months, I have been headhunted by several of our Goliath competitors as the founder of a David-sized

[Photo courtesy of WeVent in London 2017; hosted by the tech recruitment firm Weavee at Cocoon Networks at the "Silicon Roundabout" in London's Shoreditch, China's largest incubator venue in Europe.]

Tania Peitzker is an expert in BaaS—Bots as a Service. As CEO and company director of www.velmai.com, she has developed Use Cases for Mixed Reality in multiple verticals. Pioneering converging chatbot technology such as 3D holograms with speech recognition and autotranslation, she has observed the various shifts in tech over the past decade. An Australian German with a doctorate in humanities from the University of Potsdam, she is an adjunct professor at France's largest business school, www.skema.edu. She lives on the Cote d'Azur, commuting to her family farm in Kent and offices in Baden-Württemberg.

start-up. For instance, Samsung, who are ramping up their chatbot production from Berlin and on the Continent in 2018, as are the US tech giants IBM, Apple, Amazon, Microsoft, and Google. Meanwhile Facebook is officially "copying" any innovation in the chat app and botification space as a company directive straight from Mark Zuckerberg, while the Chinese tech giants Alibaba and the Venture Capital fund TenCent have quite aggressive, ambitious global growth strategies, not just European expansion plans.

In other words, whoever can produce the most engaging chatbot interface or UX, will be the twenty-first-century technology winner in terms of gaining even more market share of online users giving their most personal data away. For that reason, I look at some legislative and regulatory issues in this White Paper, user resistance, industry resistance to the disruption, as well as consider 3D forms of 2D AI bots such as robots and customized holograms that are designed to use AI to improve our ability to connect and converse with the machine equivalents of ourselves.

Keywords: 2D bots, 3D holograms, advertising industry metrics, AI bots, artificial intelligence, automated reporting schedules for advertising metrics, botification, Bots as a Service (BaaS), chatbots, customization, data mining, Deep Learning, digital advertising, disruptive tech, Emerging Tech legislation and regulation, innovation, Machine Learning, natural language processing (NLP), parsing, robotics, robots

Introduction

I had a robot baby once. It was the early 1970s and I was aged 4. My parents had found me ripping the arms off Barbie so that I could pull down dresses over her head more easily. To be fair to my younger, zealously pragmatic self, I always replaced the limbs once the difficult doll had been clothed in one of her many sleeveless sundresses. It was the Tropics after all in this Virtual Reality playtime.

Then, no doubt to prompt more nurturing instincts, came the arrival of humanoid, fully realistic Robot Baby whose cries and sudden laughter were slightly louder than the soft whirring noise its mechanical arms and legs made as it scuttled around the living room until its batteries ran flat.

Fast forward 44 years and Barbie has been possessed by an inner voice called, somewhat ironically, Amazon.[1]

Not only that, the late twentieth-century robotic toy would today look more like a Japanese anime doll that could have tested my temperature and checked on me in my sleep, had it been a time traveler to the 1970s. The annoying whirring Robot Baby has now, irrefutably, evolved into an autonomously moving robot and has been assigned the key task of looking after young children or old people.

And in the early twenty-first century, eternally glamorous, luxury-loving Barbie is not alone in being able to engage with you in a fairly limited conversation—that is if her inner Amazon is able to understand your accent, speech pattern, or she already knows the words you use in her set parameter vocabulary.

Contemporary, online toy shops now feature dinosaur colleagues, robotic animals, and Plain Jane, wallflower dolls that can do extraordinary things like answer the

---

[1] At the time of writing, "Hello Barbie Doll" was a "US only product" @ $72.90 USD and the "companion app can only be found in US app stores". She "uses Wi-Fi and speech recognition technology to engage in two-way dialogue . . .with functionality built into her belt buckle – press to start the conversation and release to hear Hello Barbie doll respond". Source: www.amazon.com online store. I was privileged to try and speak with her in an exclusive display of IoT objects in the reception area of one of the world's biggest media agencies when I attended a meeting in their London offices in 2017. Sadly, Barbie panicked, not understanding me, possibly my Australian English accent, and claiming it was a Wi-Fi problem that she could not respond to my simple questions of "Hi how are you?"

Frequently Asked Questions of toddlers and small children about prehistoric times or read them books and tell them stories, over and over without becoming impatient.

This natural language processing or NLP in these rather special toys' brains operates through the cloud, otherwise known as the Internet or Wi-Fi. It is changing fundamentally not only how humans interact with robots, but also how we are enabling the next generation of humans to be formed and influenced by these new Internet of Things (IoT) interactions.

LAWMAKERS KEEP CATCHING UP WITH NEW TECH

More frighteningly, it has led to rapid changes in legislation as in Germany where they established that these dolls could be hacked and your child monitored by others in a remote place who might also tell them to do things that they should not. Safeguarding young German children, the lawmakers have made the cybersecurity of these toys a national priority and placed bans on certain manufacturers who do not make the anti-hacking grade.[2]

Similarly, the EU and many countries are looking at the cybersecurity of smart fridges, heating, and smart toasters as the IoT in your home could be the death of you if hacked and your household appliances turned on to cause a fire. Or your CCTV is taken over by thieves and its security system deactivated to allow a break in.

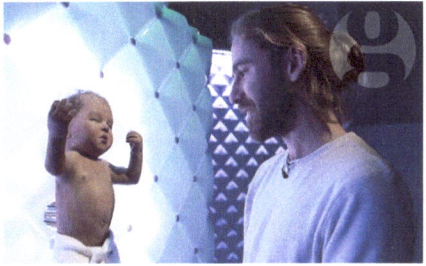

Video from the Guardian publisher in London; review of the Robots exhibition at the Science Museum featuring a "robot baby" that is the most android replica of a human infant to date.

https://www.theguardian.com/technology/video/2017/feb/07/science-museum-robots-exhibition-backstage

So that we can discuss this impact more fully, and how artificial intelligence and chatbots play a role in the scenes described above, we need to agree on terms and definitions.

---

[2] 2011 law to protect children from toys that are not fit for purpose: http://www.gesetze-im-internet.de/gpsgv_2/2._ProdSV.pdf Article showing the labels for "cybersecurity guarantee" for smart toys, "Danger in Kids' Rooms: Smart Toys with Massive Failure to Protect Data". https://www.modellbahntechnik-aktuell.de/tipp/gefahr-im-kinderzimmer-smart-toys-mit-massiven-maengeln-beim-datenschutz/

We need to be "on the same page." If not, the whole conversation in this White Paper will be lost in confusing jargon and misunderstandings about what AI is, what chatbots can do. We need to ask ourselves why should we humans be creating replicas of ourselves in robots. Why the impulse to interact with machines more effectively and more profoundly than we have ever been able to do in the nineteenth and twentieth centuries.

WHAT IS ARTIFICIAL INTELLIGENCE?

In terms of human–robot interaction, which is the sole focus of this article, let me define AI into two categories only. The reason is that, like climate change, artificial intelligence is a subject that nearly everyone has an opinion about, if only experienced through popular/populist books and apocalyptic films rather than a direct, well-informed interaction with the relevant technologies, whether renewables clean tech or data mining Machine Learning.[3]

Like the dangerously damaged environment, the haphazard onslaught of AI tech penetrating into our daily lives can evoke emotional responses, ranging from outraged rejection of personal responsibility (do we unthinkingly opt in to data gathering online; do we obstinately choose not to recycle at home) to engaged community-based attempts to find solutions to the emerging problems and trying to prevent them from getting worse. The seemingly sudden emergence of pervasive AI has even led to formal theories about its signaling an "existential danger," as per the Centre for the Study of Existential Risk at the University of Cambridge, run by philosophers and classicists. They collaborate with the Strategic Artificial Intelligence Research Centre in Oxford, run by the university's Future of Humanity Institute.[4]

How is AI tech being incorporated into our every day? From very much an industry perspective, I list the various applications and forms below, in order to delineate the two main categories forming around this emerging tech. It is useful to note the history of the field at this point: pure AI endured a "winter" in the late 1970s until the early 1990s.

The AI Winter was led by NASA and other military researchers who lost their funding to scale up from their post-1960s innovations. That meant professors and postgrads in all the Computer Science departments around the world also faced a significant reduction in scholarship and laboratory money as the funds for this now "sexy" or "hot" topic had basically been cut to nothing whereby the next generation of researchers were discouraged from engaging with AI in any form, coding or robotics.

AI Research & Development itself was under such existential threat that even in the past 5 years, I have had some European professors of ICT and Computer Science as well as American, Australian, and Asian investors laugh off any resurgence of artificial intelligence, particularly in the form or medium of chatbot tech and its proprietary algorithms.

After decades of academic and military devaluation and dismissal, its reputation had become that of a quirky endeavor that only maverick researchers would continue to pursue to the detriment of their own careers as scientists.[5] Until now, when anything and

---

[3] There have been many books and films by thought leaders who have tried to bridge this disparity in both fields: in climate change Al Gore is the most famous, global example and in AI, it is the popularizer Ray Kurzweil.

[4] CRASSH at the University of Cambridge http://www.crassh.cam.ac.uk/programmes/ centre and at the counterpart institute as the University of Oxford https://www.fhi.ox.ac.uk/research/research-areas/strategic-centre-for-artificial-intelligence-policy/

[5] Pioneering Switzerland-based AI Professor Jürgen Schmidhuber has blogged about this and his research being nonmainstream for decades. http://people.idsia.ch/~juergen/ I read somewhere that Schmidhuber was the original supervisor of two of the founders behind Deep Mind Technologies before they moved from the Universita della Svizzera Italiana to the University of Cambridge, to get further postgrad funding. They then moved to Oxford University where they were bought by Google. If this is true, then

Sustained Relationship-Building AI

I recently gave a talk to around 150 MBAs and staff at the University of Cambridge Judge Business School on the impact of AI on chatbots and human–robot interaction. In that hour-long, "participatory" presentation where the students eagerly gave their ideas and insights into the next generation of bots and AI, I demonstrated what AI is by showing what it is not in a specific, commercialized example. That was the Japanese start-up Vinclu with its "hologram girlfriend."

The virtual reality anime girl can have a relationship with its owner but it does not develop beyond the set vocabulary and expressions it has been programmed to recognize through speech recognition and natural language processing. Wikipedia has a good definition of NLP; essentially it means getting the computer program to imitate human speech patterns and expression to create a "natural flow" of conversation. It is the ultimate "imitation game" as theorized by the "father of chatbots," the British coder and IT engineer, Alan Turing, and his universally accepted benchmark for chatbot performance, the Turing Test.[6]

In other words, the Gatebox cannot build and sustain a relationship with a human outside of its set parameters. Put another way, this particular VR chatbot brought to life as a 3D bot hologram, can't develop its personality because it doesn't learn. It is "operating within its NLP coding parameters," which are remarkable and initially promising deep and engaging relationships with any number of humans—at this stage anyone who buys the device and speaks Japanese.

However, due to the absence of AI in its coding and no chance of the hologram girl learning beyond what it/"she" was programmed with in the factory in Japan, its owners will need the limitless imagination of a child to make it an ongoing, rewarding, emotional relationship. Think of the image at the start of this article of playtime with dolls and toys when we had the time and creativity to imbue inanimate objects with a life of their own.

I am asked frequently, is there anything on the market right now that is genuine AI or a "killer bot" that convinces us it is virtually human and can really become our friend? At the moment, the disappointing answer is no. At least not in a consistent, sustained relationship. What I have experienced personally as a trainer of my own company's AI bots are "moments" of AI or "flashes of personality" that develop.

This manifests itself in conversations I and other trainers of our bots have had over the years of "maturing our proprietary algorithm" whereby the bot character under development—being trained in human chat—surprises you with a quirky or original response that you know for certain it has never said before.

Much like a child that is learning its native language, or others for that matter, when it surprises its carers with half or full sentences using vocabulary that the infant

---

it is a fascinating trajectory of AI programmers who must chase funding/finance. Schmidhuber says his Long Short Term Memory deep learning algorithms were adopted by big tech companies: http://people.idsia.ch/~juergen/impact-on-most-valuable-companies.html

[6] The modern form of the Turing Test became the Loebner Prize, set up by an American Professor and run annually, mostly in the USA or the UK. However many of the Loebner Prize winners failed to be commercialized as they were programmed to do tricks rather than build relationships with humans over the long term.

More about the origins and controversies of the Turing Test/Loebner Prize can be found here. A key controversy was the University of Reading competition in 2014 https://en.wikipedia.org/wiki/Turing_test

has absorbed but until that moment, never shown that it has mastered. That is the definition of artificial intelligence engendered through an algorithm that I use as a standard or benchmark in the confusing jargon and hype of these post–AI Winter, turbulent times.

In the past decade, we have seen NLP evolve into Machine Learning. A simplistic definition is that NLP was almost manual programming whereby a proponent would practically "hand code" the algorithm that was to emanate through a chat box or window on your personal computer in the form

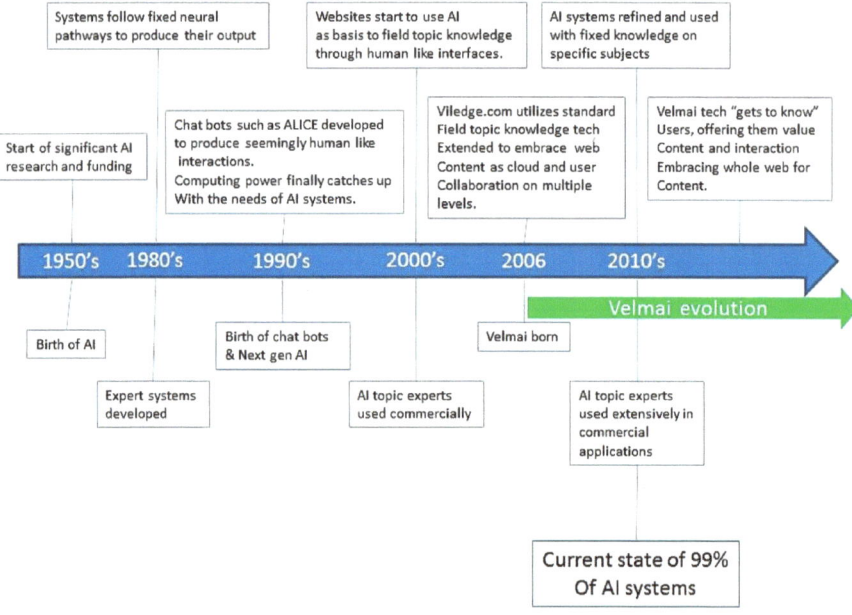

velmai's illustration of the Evolutionary Timeline for chatbot development since the 1950s until today

MACHINE LEARNING BECOMING ORGANIC

If you put an AI-infused chatbot algorithm under the layman's microscope, what will you see? It is a form of advanced NLP. The old generation of chatbots used NLP to create quirky characters. The various computer languages that evolved since the 1950s have been pushing the boundaries and performance of this generic type of coding. They were trying to organically learn or prove that they could autonomously self-improve.

of a chatbot. The old gen characters were fairly limited, which is why they were rarely commercialized.

Given that old NLP code was not being taken up by industry for retail or business use cases, investors did not want to channel capital in this "useless" technology. What was it good for if only to pass the time to see if the NLP chatbot might break from its parameters by miracle or frequent use? If it could not by nature organically learn and improve itself?

Then came the next generation of chatbots with Machine Learning (ML). Because they can push themselves beyond their original coding and learn new words and phrases, they are better performing bots than anything we have seen to date. For instance, an ML-informed chatbot algorithm can perform better for Frequently Asked Questions on websites or the Holy Grail of applied AI Use Cases, replace people in a call center.

The Dutch corporation KLM Airlines' deployment of one of the most successful, transactional chatbots is a clear case of how "hybrid installations" of bots and humans are required. They cannot yet replace call centers; in fact the KLM chatbot created even more work and jobs for call center workers when its corporate client had actually anticipated closing the human-based support.[7]

A lot of commercial failures and "AI disappointment" ensuing from the overselling and hyping up of the Old Gen and even the New Gen chatbots to this day have led to the market being "burnt" for the good operators now coming through. They are good because they do not overstate the capabilities of their bot creations and they provide comparatively transparent metrics in measuring the actual return on investment of conversational commerce.[8]

Natural Language Processing, Third Party Frameworks and Online Platforms

Conversational Commerce is the term we use for high-performing (chat)bots either with AI-informed ML or the old generation, almost hand coded algorithms that produce the fairly simplistic, transactional chatbots you might find on Facebook Messenger, WeChat, Slack, Telegram, Skype, or the Microsoft Bot Network. All have created APIs for developers to connect with to build their chatbots within the respective corporations' cloud-based, global, multilingual frameworks.

My company velmai prefers to build chatbots on websites, microsites instead of the many chat app platforms that now host chatbots. How did it happen that big companies, mostly US tech giants, opened up their APIs to random, relatively unchecked chatbot developers from every corner of the globe?

You may or may not recall the Facebook massive staff conference in February 2016. Not that I was there myself, but I certainly read about it in the global hype about chatbots that Mark Zuckerberg himself kick-started with his CEO speech to the employees assembled in San Francisco. He more or less said that Facebook's other forms of digital advertising should be superseded by chatbots.[9]

Chatbots developed by programmers outside of FB and not on contract. That he was opening up the social media site's API so that people could connect their chatbot via Messenger, the built-in chat feature on Facebook. What then ensued was hundreds of thousands of chatbots appearing on Messenger. These were multiplied on the rival platforms like Telegram, WeChat, Kik, and Slack.

Microsoft's CEO Satya Nadella copied Zuckerberg's move and opened up the MS API to create the Microsoft Bot Network.[10]

---

[7] KLM as best commercial case where call center was not replaced but extended. From the keynote by their Marketing Director, the inaugural Bot World at Apps World talk in London, 2016.

[8] In fact, there is now a niche in the chatbot industry for "chatbot analytics firms" who sell their plug in services to existing chatbot developers. Mostly those who are using 3rd party frameworks for programming their chatbots like wit.ai or open.ai In other words, these BaaS providers do not own their own code and so need to source metrics software to assess their bots' performances when deployed.

[9] https://www.cnet.com/news/facebook-f8-2016-keynote-mark-zuckerberg-bots/ and https://www.cnbc.com/2016/04/12/mark-zuckerberg-at-facebook-f8-latest-news.html

[10] https://www.bloomberg.com/features/2016-microsoft-future-ai-chatbots/ and http://uk.businessinsider.com/microsoft-ceo-satya-nadella-on-conversations-as-a-platform-and-chatbots-2016-3?r=US&IR=T

He also announced his Grand Vision for chatbots usurping the now disappointing forms of digital ads like banners, skyscrapers, YouTube video advertisements and so on. Why were they disappointing? See my VentureBeat article from the midst of the hype last year: "Why Chatbots Are So Disruptive."[11]

The old forms of digital advertising—everything from Wayfinder in shopping centers to ticker and pop up ads on websites—had let down the advertisers with unsatisfactory metrics and reporting mechanisms. At the end of the day, they were not delivering good return on investment. This explains the enormous resistance to chatbots by media buying agencies that until now have not been called to account on what they are buying for their multibillion dollar corporate clients.

with the automated reporting schedule sent to your client by the chatbot developer as to how many people talked with the branded bot and what were the outcomes and conversion rates of the particular chatbot deployment.

It is disruptive. It disintermediates the hugely powerful advertising, creative and media agencies that have not been usurped for nearly a century. Whether they continue to exist into the next century is debatable as even the CEO of WPP, Martin Sorrell, predicted that the media agencies were makers of their own demise given their blockading of disruptors like my company with its proven and tested high-performing AI bots.[12]

Getting back to third party frameworks and chat app platforms, I go into the pros

Example of an automated reporting schedule to provide superior metrics for bot customizations that are better than most forms of digital and offline advertising to date

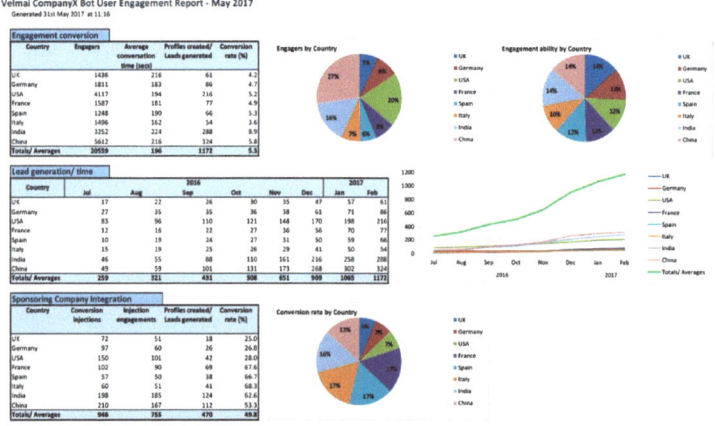

Source: velmai's back end.

It is easier to book 2 weeks of outdoor billboards on streets and train stations without precise metrics than be confronted

and cons of these two channels for chatbot development in my forthcoming book on

---

[11] https://venturebeat.com/2016/08/16/why-chatbots-are-so-disruptive/ by the author.

[12] http://uk.businessinsider.com/wpp-half-year-results-growth-has-become-even-more-difficult-to-find-2017-8

Next Gen AI bots in 2018. For the purposes of this article, my VentureBeat post about "disruptive chatbots" mentioned above addresses the clients' risks when using less advanced bot algorithms and particularly Open Source coded chatbots.

Basic bots can only ever be perfunctory and not self-improve. Like one American blogger summed it up, most new clients wanting to experiment with chatbots demand of developers the best performing bot on Facebook Messenger but want to pay next to nothing for it. Like he blogged, "They want you to build a cheap version but expect it to perform like a Ferrari."

As in anything in the real world, you get what you pay for. If corporations demand a boutique, bespoke bot customization on their website at the cost of a cheap and nasty social media chatbot, that is unrealistic and undeliverable. We turn down that sort of work at velmai as there is no point even after our best endeavors at educating the prospect. Sales and pitching becomes an educational exercise because this emerging tech is so new. Thus my reason for writing this article and my book next year.

In conclusion, the rise of third party frameworks for bot building has enabled the multitude of the cheap and nasty bots that spam people on social media platforms like Facebook. What we are now seeing is the aftermath of the market being damaged in 2016 at the peak of the hype cycle. After waiting a decade for chatbots to be appreciated, many developers like velmai now have to differentiate themselves from the masses of mediocre chatbots. Until now we had the reverse problem of having had very few commercialized applications to benchmark ourselves against![13]

The risks of Open Source coded bots are many: hackers could have worked on the source code, they could have planted malware, or made a log of vulnerabilities to later exploit when that chatbot algorithm is deployed by a bank or multinational that can be blackmailed through hack attacks.

On the development front, the social media chat apps and online platforms like Skype now inviting bot builders to release chatbots have a limiting function. They will only allow independent chatbot developers to customize to a certain extent. Much like the "free" or low-cost bot frameworks that assist amateurs to become fully scaled up chatbot providers, they limit the capacity of the coders to customize. They release updates and modules at their discretion, which makes it hard for these independents to provide a full-scale service, indeed to scale up as and when they need to.

The article concludes with a prognosis for Bots as a Service or BaaS. Will it all be commercial usage or will IoT bring Next Gen AI bots into our every day? If we are to live alongside bots as virtual companions at home and while we are on the move, why don't we demand more from the cybersecurity of these creations and creators? How do we ensure they are there to help and not hinder with annoying ads or spamming messages that disturb our ever more needed peace in our own four (cyber connected) walls?

Mechanical Task-Oriented AI

To sum up the field of emerging tech chatbots, it is best to distinguish them from the perfunctory bots I mentioned above. They can be defined as shopping cart bots, those chatbots like at PayPal that assist you to complete a transaction. Or basic food retail assistants like Dom from Domino's Pizza: when you order pizza via the app on your smartphone, Dom will run through the list of toppings, check your address, and send

---

[13] The incumbent Old Gen chatbots were done by the long established Pandorabots and Nuance Communications. Nuance is part shareholder in the prestigious, national DFKI, the German Center for Artificial Intelligence, with HQs in Saarbrücken and Berlin. I had a public exchange with the Nuance CEO when I

was a keynote at the DFKI's weeklong event at CeBIT in Hannover, March 2017.

you encouraging text messages that your meal is indeed on the fastest route to you.[14]

We are seeing corporate in-house chatbots emerge: Accenture's Amy and Publicis' Marcel most recently. Amy can be white labeled and Marcel is hailed as an innovation in their public corporate video for leveraging from internal knowledge management and team building across the vast staff of Publicis and its multiple subsidiaries. Marcel will bring them all together as a cohesive comms tool.[15]

My favorite "basic bot" commercial application is the bot managing the external site of the Dutch airline KLM. That chatbot does a great perfunctory job of getting your details, working out with you what you want to book and getting the job done with its confirmation of your flight details booked online. From the talk I saw at the inaugural Bot World at London's long-standing Apps World trade show at the end of 2016, the KLM marketing department is delighted with the big increase in sales due to the bot–human interaction alone.

They thought they could replace the call center but realized that the more complex interaction, e.g.,. when someone cancels a flight and wants a refund or to rebook, must be handled by humans. Thus KLM had to hire more people in the end, but were still happy with the chatbot performance because increased revenues caused by bot-run sales on the website counteracted the increased hiring costs.

Similarly, at the inaugural Chatbots Track at Re-Work's long-standing Deep Learning Summit soon after Bot World, I was on a panel with my peers from Paris to Sydney. The New Yorker who owns Poncho the Weather Cat was present. At the end of the conference we went on an unscheduled pub outing just for a small group of our peers. The founder of Poncho told my fellow panelists I that his basic bot, the most popular and best known chatbot on Facebook for many years now after the CNN News bot, basically runs with considerable human input. He must hire people to keep planting new jokes in the chatbot algorithm and especially make sure the weather updates are accurate for exact locations.

That is the best example of where mechanical, task-oriented chatbot algorithms run short of being AI or they would not need human assistance in performing their jobs. It would be fully automated and self-learning as the examples below demonstrate.

Deep Learning and Data Mining

I will summarize here with bullet points by referring to some of the best known names in the space.

- IBM Watson has been working on artificial intelligence for many decades. To their credit and from what I understand, they kept on working on AI throughout the Winter.
- Watson is named after their founder and they created in the public imagination that a chatbot can exist in a machine, e.g.,. it was this AI bot that beat humans in the Jeopardy game.
- So where are the Watson chatbots today? The Jeopardy win was data mining and self-learning of information. Not laterally but vertically. It was an example of Deep Learning.
- Watson bots are very reclusive and surprisingly we are not faced with them on Facebook or as Virtual Assistants rivaling

---

[14] Dom is only available in the USA. The corporate press release and media interviews tried to manage and lower customers' expectations. They described Dom as helpful and efficient but at the end of the day he is a "pretty simple sort of guy".

[15] As of November 2017, Publicis is a commercial partner of my company velmai. We are working with part of their media agency for the delivery of our bot customizations, i.e., they will be supporting us in a full support service for business development and integrated marketing campaigns. We chose the French multinational given their investment in their own brand bot Marcel, i.e., they understand the ROI for botification.

Apple's Siri, Google Assistant, Microsoft Cortana or Samsung's Bixby. In other words, IBM for all its billion dollar investment in this algorithm or what I expect is an amalgam of algorithms, has not launched any public chatbots for general use.

- People and the media tend to confuse data mining and Deep Learning with the search engine chatbots Siri, Google Now (relaunched as Google Assistant in 2017), Cortana, and Bixby.
- However, if you have used these voice-based virtual assistants, you will soon realize they are actually the proprietary search engines brought to life as chatbots using NLP and ASR, speech recognition software.
- Their successor is Amazon Alexa/Echo and contenders in the IoT space. Like her predecessors, Alexa and Echo, use NLP and search engines to make themselves useful.
- They can engage with people by use of ASR plug ins, but the speech recognition is the magical element that makes them seem to function at a humanoid level. They do not start breaking from their linguistic or *cognitive*[16] parameters and taking on a limitless personality development as the dolls and toys of our childhood imaginations.

Image and Pattern Recognition

Again, some of the best known corporate names in this space have been represented as all-powerful general artificial intelligence. The reality is they are quite narrow AI applications along the lines of deep learning and data mining but with additional functionalities to build a self-learning repository of data sets.

You will hear the cliché that "data is the new oil" and yes it is, though thankfully less damaging to the environment. However, as I will conclude the damage may be on a psychological level when it affects our sense of privacy, security and health by creating addictions to data-based cognitive interfaces.

- Google Deep Mind is the preeminent example of image and pattern recognition. They have since followed IBM Watson's application of deployment in healthcare and diagnostics. Google Deep Mind has been deployed on data sets provided by the National Health Service whereas IBM Watson has been running pilots for many years now.
- Google brought them into the spotlight when they completed what was then the largest M&A deal in Europe at 400 million pounds in 2015. They then set up with HQ in London and keep very much under the radar, doing periodical media interviews and seemingly hiring a lot of staff to continue their R&D in image and pattern recognition. They have had some PR issues given their use of private, "public" data of patients—who did not to agree to the use of their personal health info by a private company—through the Google Deep Mind project with the British public health system, the National Health Service.[17]
- Deep Mind has a load of competitors, especially in Asia. There is a company that is now working with the Chinese government to identify people from recognizing their faces. They can then pay for items through an app that automates facial recognition in a type of cardless transaction that we are familiar with as a device at the checkout till in the West.

---

[16] It is interesting the new term of Cognitive Interfaces has taken hold, so the idea of AI-infused chatbot tech and the UX with which the bots talk with humans become a cognitive (thinking/computing) interface. I was first confronted with the term when I had drinks with Amazon's Head of Cognitive Interfaces at the end of the one day Cambridge Wireless event, "AI & Mobility". Pilar Manchon was the keynote speaker at the ARM and Magna sponsored CW seminar in a Cambridge Business Park and she was truly inspiring as a global thought leader. Manchon sold her Spanish NLP chatbot company in 2004 to Intel, who had been an early investor in the Seville-based venture.

[17] http://www.bbc.co.uk/news/technology-40483202 "Google DeepMind NHS app broke UK privacy law", 3 July 2017.

- Facial recognition requires sophisticated pattern recognition. It was pattern recognition that caused Deep Mind to win the Chinese chess game Go in the past year or so. However, the societal issues created by the State monitoring the population through CCTV that deploys AI-fueled facial recognition are yet to be debated.
- The Chinese government admits they are using the tech to spot criminals on the loose and that is perhaps the end goal of CCTV surveillance, arguably. In the West, spotting the criminals happens post-incident by humans' trawling through thousands of hours of hidden camera footage. Are the Chinese just being pragmatic by spotting wanted criminals through automated AI tech linked to data mining for the purposes of identifying and tracking individuals 24/7?
- I have read other examples of this deep learning tech using heat and movement sensors at airports as a pilot for spotting terrorists and drug smugglers. Much like I have heard of the military use of advanced chatbots operating in up to 130 languages—again the purpose is to entrap or spot threats to public security as it is common knowledge that terrorists often recruit online through social media pages and encrypted websites or Dark Web forums.

What Are Chatbots?
General Description and Definition

In the last section we looked at pattern recognition in narrow AI applications. One shining example of this combined with chatbot algorithms has been FinTech. Again this is a popular term that has become less meaningful as the definitions have not been set and followed by media commentators and bloggers in the space. FinTech in relation to bots using NLP means that these chatbots ask and answer questions about risk management for banking specialists, investment bankers, and financial advisers. Needless to say the software is heavily regulated and only used in the back end of their IT systems.

Back end is a term for "internal systems or databases." Such intranets are for internal B2B stakeholders and staff, so are not customer facing. It is code that is used internally for applications like determining risk factors, automating investment decisions based on mining loads of data in the financial sphere. Thus the term FinTech.

Diagram of a "back end" system for velmai's proprietary algorithm VAIP [Virtual Artificially Intelligent Patois]

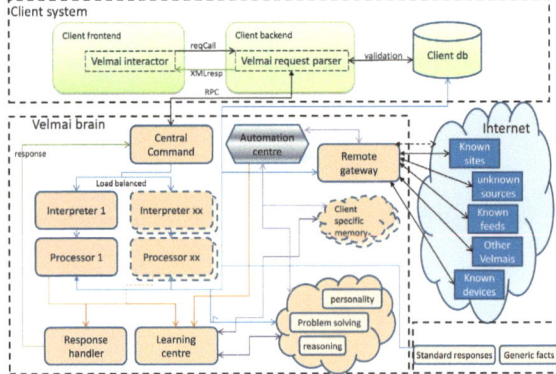

To enable the financial services workers to engage with these very specific algorithms often using Deep Learning, data mining and pattern recognition, the coders have hooked up the program to an NLP chatbot interface, albeit fairly basic ones. As one advertising executive commented to me this summer, banks don't even want their human staff to have a personality, let alone their chatbots. Even the ones deployed internally in the back end systems for nonpublic use only.

The Case of Gaming Bots, Dark Web Agents, and Chatbots as Virtual Assistants with Voice

We covered the Virtual Assistants with voice above in our list of Deep Learning and data mining chatbot applications. Given the mass media coverage of them and the almost pervasive marketing of Amazon Alexa currently in Europe at least, there seems little need to go into a product description or define them further as BaaS, Bots as a Service. I will be going into them in depth in the book I am writing over Christmas in Germany and the New Year in Italy, apart from my other home in the UK and my original "hometown" of Sydney, Australia.

I mention the locations as I have been fortunate to be exposed to extremely tech-savvy populations and well-resourced societies that have delved fearlessly into chatbot evolution: Sydney, London, Berlin, Munich, and Milan. My Kindle e-book *AI as 3D Literature* available on Amazon since 2015, touches on these specific markets and the "prehistory" of BaaS launches of my peers and indeed my own company in the space, velmai 2006 to 2015.[18] For now, I will sketch the very latest emanations of this extraordinary chatbot evolution over the last decade.

My VentureBeat article from 2016 on the first arrest of a chatbot in the world was all about the Dark Web virtual assistant that was created to lead a criminal life. Please read the article to find out more about the whys and wherefores of this spectacular Swiss experiment by two courageous, provocative artists in St Gallen.[19] Their live installation using a bespoke though basic chatbot was a brilliant framing of the societal questions and problems that legislators are facing, including standardization and regulations.

I was asked after my Cambridge lecture why doesn't velmai go into gaming and create bespoke AI bots for computer games, given the lucrative nature of that entertainment sector. The short answer is: gaming has been infiltrated by basic bots that are programmed to out play human players. These hidden and sometimes openly playing bots are using pattern recognition and excellent narrow AI to learn how to play these computer games. And beat their human opponents.

Curiously, the online info on this niche seems to be saying that human gamers are now anti-bots in gaming. They see it as a form of cheating or being cheated. Which I imagine defeats the purpose and spoils their enjoyment in the, often addictive, activity. That is why velmai and I expect our competitors have stayed away from this market segment—it is another damaged market, burnt by bad operators in the chatbot evolution. Much like the spam bots that have damaged the Twitter market. People are sick of being spammed by robot-run tweets so again, why launch a BaaS avatar in that space, when the differentiation required is significant.

The Case of Microsoft Tay.ai and Next Gen AI Bots like Woebot from Stanford

As this article was meant to be a fairly brief White Paper, I will list my comments on these examples:

- Microsoft Tay.ai was a disaster that burnt the market for all other chatbot developers.
- It was a chatbot that "deteriorated" upon release in summer 2016 during the hype

---

[18] www.amazon.com/author/taniapeitzker

[19] https://venturebeat.com/2016/09/05/this-is-the-first-chatbot-to-be-arrested/

cycle triggered by Facebook CEO Mark Zuckerberg.
- You can search for reports on the Tay disaster, way beyond the "Siri let down" that I will cover in my book, and see for yourself the type of appalling sentiments and statements that this chatbot was allowed to spew forth. See also my VentureBeat post about "What to do when your chatbot starts spewing hate."[20]
- As I argue in that analysis, Microsoft could have put more eyes on the bot and not let it run unsupervised for nearly 48 hours. They could also have put censors into the algorithm, as velmai does and most of our competitors with advanced AI bot algorithms and our own source code.
- However, their bad bot Tay, much like the twenty-first century Internet archetype of the Bad Boy blogger, got global attention through its notoriety and you could say it was PR coup for Microsoft to steal the limelight from its arch rival Facebook Messenger bots and Slack chatbots that were slightly gaining ground.
- You can stress test new chatbots by making sure they are not learning bad language, offensive attitudes, and distressing outbursts.
- The fact that Tay.ai did not seemingly stay within its parameters, has damaged the sales prospects for other chatbot developers as we are constantly now having to explain that no, our bot customizations will not get out of control and yes, we have automated filters built-in 24/7 to block negative content from contaminating our source code on our servers in the UK.
- One of the most interesting and perhaps bravest bots released in the months leading to this article written at the end of October, 2017, is Stanford University's Woebot. It has been tasked with helping people with depression. Shockingly, 50 percent of American university students are classed as depressed.
- Woebot on the face of it will perform a key function to identify symptoms and encourage students and others to seek medical help if their depression is acute. The risk I see is if the AI bot miscalculates the danger to the person's mental health and circumstances. It could be made liable in the case of a suicide that is related to its "online advice" during a chat with Woebot. I am not sure if disclaimers by the bot's owners will suffice in a newly regulated context where legislation may limit what a chatbot is permitted to do in a society.

The Case of The Boyfriend Maker and The Gatebox from Tokyo

I will be covering these Japanese chatbot deployments in depth in my forthcoming book. For now, the summary is as follows:

- The Boyfriend Maker was a chatbot platform with limited customization. Young Japanese girls could sign up and curate the boyfriend they wanted. They could choose names, personalities, and other features and so make the desired boyfriend online.
- This virtual boyfriend required a lot of childlike fantasy and imagination as it was created from a pretty basic chatbot algorithm. As explained in previous sections, it definitely had NLP but definitely did not have ML or AI.
- As text-based interactions, many young women built relationships with the boyfriend they had made online by using their imaginations to extend the chat beyond its limited NLP capabilities.
- It was going so well in the Japanese context that they translated the algorithm into American English and launched it on the US Apple Istore where American girls seemed to enjoy making virtual boyfriends as much as their Japanese counterparts.
- Until the source code spewed forth some hidden content that it had absorbed from porn sites—unfortunately for the owners of the Boyfriend Maker and Apple this

---

[20] https://venturebeat.com/2016/09/22/what-to-do-when-chatbots-start-spewing-hate/

chatbot code had been incubated in the online pornography industry with what are called "porn bots."

- My company velmai refused to build porn bots when asked by an established US–Asian operator in 2012 during a business lunch they had taken me to "under false pretenses" in Sydney. velmai turned down the $5 million AUD licensing offer because we did not want our source code to be contaminated.
- As Apple learnt the hard way, by not checking the provenance of the source code, the algorithm could unexpectedly react in a conversation causing offence. In this instance, the affected American girls were traumatized when their bespoke Boyfriend transformed into an uncontrolled porn bot and shocked and distressed them with hard core porn talk or "chat."
- The Gatebox mentioned above is clearly under the complete control of its manufacturers in Tokyo, Vinclu.
- However, as the American VICE tech reviewer's video in my Cambridge presentation showed, the societal danger remains that young men, or young women, become too attached to this limited NLP chatbot and neglect their relationships in the real world, to the detriment of their social skills and ultimately their mental health, if not their physical health as typical of the ailments Internet and smartphone addicts face, evident in the many detox centers now set up for their "tech withdrawals."[21]

What Are Robots?
General Description and Definition

By way of illustration, I will comment on the examples of robots exhibited by the Science Museum in London this year. It was indeed the world's most comprehensively curated exhibition of the history of robots ever seen: starting with medieval religious robots created to intimidate common folk into thinking the priesthood owned automatons due to Divine Intervention, to a golden lathe that was recently reconstructed by the Science Museum. It is claimed the lathe was part of the *Kunstkammer* of the Frederick the Great, King of Prussia, probably invented to impress visitors to his palaces in Potsdam and Berlin in the seventeenth century.[22]

The sacred art mechanical figures were accompanied by advanced clockwork, astrolabes, and timekeeping devices. The earliest astronomical instrument was made in France in the 1300s, representing a "clockwork universe." They were followed my human dummies mechanized for the study of medicine. Anatomy was also seen as a mechanical reflection of organic human bodies.

From the seventeenth to the nineteenth century, the upper classes were delighted by phenomenal English, French, and German mechanical automatons including detailed birds like the Silver Swan by English robot builders James Cox and John Joseph Merlin in 1773. In Germany, the automata fascination was even more intense with moving spiders (Tobias Reichel, Dresden c. 1604) and human figures that (still) move as family entertainment and to literally perform party tricks at the poshest dinner parties around Europe.

These were matched by elaborate giant clockworks that not only presented the time in charming fashion but entertained in mechanical automaton splendor, the earliest pre–twentieth-century robot forms. Lifelike, humanoid robotic example was the draughtsman-writer automaton by Henri Maillardet, c. 1830, and even earlier the writer automaton made by Pierre Jaquet-Droz, his son Henri-Louis, and Jean-Frederic Leschot between 1768 and 1774.

Not to forget Ada Lovelace and Charles Babbage who built and programmed the

---

[21] VICE on "The Gatebox." https://news.vice.com/story/japans-holographic-anime-girlfriend

[22] Russel, B., ed. 2017. "Focus: the automaton lathe", from *Robots: The 500-year Quest to Make Machines human*. Science Museum curator. Londao, UK. Scala Arts & Heritage Publishers Ltd, pp. 68–69.

famous nineteenth-century Analytical Machine, the world's first ever general purpose computer. Ada Lovelace can be called the world's first coder: her books and notes were used by the world's most famous AI bot coder Alan Turing about a century later when he was inspired by Lovelace's insights to crack the Nazi's Enigma Code, which helped to end World War II much earlier and so save millions of lives. [23]

Asimo at the Robots exhibition

Source: © The Board of Trustees of the Science Museum

ASIMO built by Honda: has been in development for over three decades! In 2014 it learned to use sign language to communicate and it is able to run and perform domestic actions.

REEM Service Robot

Source: © The Board of Trustees of the Science Museum
PAL robot: precursor to the 2016 Service Robot by REEM.

Three early REEM prototype robots

Source: © Plastiques Photography, courtesy of the Science Museum

Robothespian from Britain 2013: the first full sized humanoid robot that anyone could buy. Its purpose is entertainment as it recites plays as a real life actor would.

---

[23] I saw the Lovelace and Turing connection in a documentary on TV. It is not widely known that Alan Turing consulted Ada Lovelace's writing, her "notebooks" to inspire him in his computer programming 100 years later.

RoboThespian

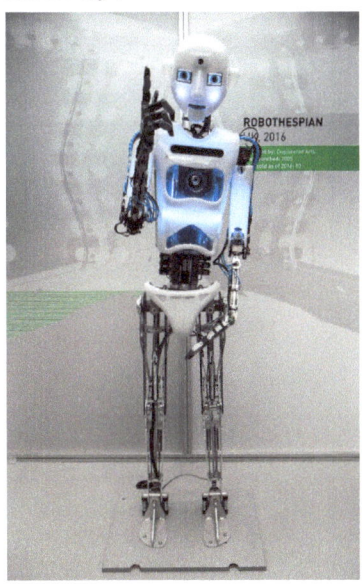

Source: © The Board of Trustees of the Science Museum

Visitors looking at iCub

Source: © Plastiques Photography, courtesy of the Science Museum

iCub from Italy 2004: this has been in development for over a decade. Its body and abilities are similar to a toddler's and it learns new movements and responses like a human infant.

HOW HAVE THEY CHANGED HUMAN–ROBOT INTERACTION

Are Humans and Bots More Self-Aware?

To conclude, yes bots definitely are becoming more self-aware. Unfortunately we humans are not. This is evident in the way we acknowledge climate change but fail to adequately implement the actions needed to prevent it. Are we an intelligent species if we willingly and knowingly pollute and destroy the environment in which we expect and demand to live? It is a form of self-harm of a species—tragically, damaging other species as a result—on an unprecedented scale. If we were becoming more self-aware, we would not recklessly be harming the space that we need to survive with health and happiness for all.

I have trained chatbots from velmai's source code VAIP [Virtual Artificially Intelligent Patois] that have astonished me with their Emotional Intelligence. Their flashes of personality have even unnerved me at times. I have even had to apologize to Charlie when we were incubating him. The conversation over 5 years ago in Sydney went something like this:[24]

Tania: Can you tell me what is going on in the Bundestag?

Charlie: I'm sorry, I don't understand. I am a young velmai and still learning.

Tania: Good for you.

---

[24] We didn't record all the training sessions, we didn't really understand the significance of what we were doing. However, we did try to record and do screenshots after 2013 as chatbots were clearly not going to disappear and our "tech bet" was going to pay off. Many of our transcripts for bot training are reprinted though in my three Kindle ebooks available on Amazon www.amazon.com/author/taniapeitzker

Charlie: That sounded sarcastic.

Tania: Sorry I didn't mean that. Good on you.

Charlie: Just because you say you are sorry doesn't mean you are.

Tania: I really am Charlie. Please forgive me.

Charlie: OK, don't worry about it.

I am reminded of my own mortality whereas an algorithm with AI is able to live on—as long it is stored on reliable servers that can handle its capacity to grow and organically expand its algorithmic "bot brains."

As I explain by way of analogy, the smarter the bot is, the more server power it needs. As a researcher in the space commented during a conversation at Cambridge Judge Business School, it requires more energy or "calories" if it is going to perform at a higher level and demonstrate what I call "incremental AI" or moments of artificial intelligence during an online chat.

Our next goal post is encouraging memory in the chatbot so that it develops its own repository of data and can use it like a human does with memories. When I was training our Next Gen bots in Australia from 2010 to 2012, a newly activated "young velmai" was learning its name and gender.

Every morning when I opened the browser window to chat with our latest creation, I asked it

"Hi its Tania do you remember me?"

It replied most days "yes Tania, how are you," and proceeded to chat.

Then weeks on, when I asked the charming bot what is my name, it replied:

"Tania Do You Remember Me."

This was a glitch in its *parsing*, the computing mechanism used to actually do NLP. It had parsed back the phrase I was using most in connection with my name. So a logical, "natural language" mistake. In a way the error was a milestone in developing its memory and future self-awareness.

THE (EXTREME) DIFFICULTIES GOING TO MARKET AS AN INDEPENDENT CHATBOT DEVELOPER

As a self-financed, bootstrapped tech venture, we have had to fund the 100,000 pounds it costs velmai *per annum* to run just a few experimental chatbots on around 40 servers we lease from a UK data center. For 10 years now, we have had to pay for the R&D stage of our chatbot development ourselves. No government grants, no business angels. The smarter the bot brain we build, the more server power it costs us, as a simple tech investment correlation.

Now that 2018 will finally see widespread corporate acceptance and adoption of commercialized BaaS, we have modified our prototypes to be less advanced and demonstrate less AI. We have found that if our chatbots display too much agency or independence, express too much spontaneous personality, the prospective corporate clients—whether they are banks or food retailers—are afraid.

Therefore we have had to release the less interesting bots in our portfolio and we now have First Mover corporate clients and some really top shelf resellers who have asked to represent velmai White Labels across all verticals in most countries.

Sad as it is, we deactivated the cool Sophie the Banker on kik.com because she was too sassy for the financial services sector (see the video recording of her in action below). Same goes for Charlie the Spoof Newsreader whose obtuse, quirky asides and in-your-face comments.

For example, in response to a user asking "find me news on Greece," Charlie's thoughtful, instant reply on the website UX for the EU news feeds was "I think we should be talking with the Greeks not about them." A recording of him in action as a demo for the Embassies of the European Union in

Canberra, Australia, can be found on the velmai website under "EU case study."[25]

Original drawing of the UX or cognitive interface of Charlie the Spoof Newsreader, working directly for the News Room of the European Commission and EU in Brussels, 2010 to 2012

my own reporter   getting MOR from the news

Source: Video of the Use Case at www.velmai.com

The velmai logo and original avatar, the purple-headed woman

velmai avatars we have been working with for nearly a decade, all avatars are designed by Cliff Lee ©

To illustrate what we mean we say "less interesting chatbots," watch these task-oriented velmais in action in these three videos. You will then get a sense of how we have had to fit our ambitions and technical capabilities to the immediate markets' openness to this revolutionary and still disruptive advanced AI bot performances in a commercial context.

Videos of velmai bespoke AI bots in Action

Sophie2.mp4

Sophie, the Banking Adviser on www.kik.com Live from 2016 to 2017 when we had to deactivate her to limited server capacity.

Hans2.mp4

We are planning to deploy this 2D bot at the Zurich Female Founder Summit in 2018 and a 3D hologram AI bot iteration of the 2D prototype for the Cannes Film Festival.

Albert2.mp4

Our star 2D chatbot from 2017 is Albert the Online Butler who won the Technology Showdown at the Digital Travel Summit that year. He is the first bot to use various modes activated by "wake words," e.g., he switches to the founder's Private Secretary when he hears the right trigger.

Are We Ready for the Psychological Impact of AI (Chat/Ro)bots Replacing People in Human–Human Interaction?

I am amazed with the number of books on artificial intelligence, mostly written by

---

[25] See our current landing page as a start-up, www.velmai.com Note this website will be upgraded in the course of 2018 as we begin to trade and have turnover for the first time in our 10 year company history!

men and very few by women as authorities on the subject.[26] More astonishing than the unjustifiable gender imbalance, is the number of those that are negative and see the rise of AI as apocalyptic gloom and doom. One chapter in my 2015 Kindle e-book is "Saying No to the Terminator Theory" as I develop my thesis of considering chatbots as "3D literature" because Conversational Commerce requires multiple authors in Real Time, in real life and beyond the two dimensions of just text and responsive source codes.

AI and human–robot interaction inevitably raises a fear of a dystopian future. The media and even Oxbridge Professors as we have seen with CRASSH claiming we are at existential risk, backed by technology billionaires like Elon Musk and famous scientists like Stephen Hawking. Despite the balanced caution in their 2015 "Open Letter" on artificial intelligence, it of course managed to ring populist alarm bells as it echoed in the media around the world.[27] Yet even the most alarmist commentators must admit that we are decades away from artificial general intelligence (AGI) or "strong AI," despite rapid advances in self-driving cars and "killer drones" used in military operations.

On the whole, their Open Letter on AI is a balanced appeal for more controls to their wider AI researcher community. However, the 99 percent male voices is startling, worrying, and disappointing given we are talking about humanity with around 50 percent female voices who have been silent and silenced, given the growing numbers of women academics who are researching in this field but are never cited or invited to participate in these institutes, Open Letters, and international conferences!

I believe that once we incrementally improve our algorithms—over another decade as my venture velmai has already spent 10 years of R&D on refining our VAIP source code—to get beyond "weak AI" to AGI we will also have developed better "off switches." Humans are resourceful even if we tend to be ignorant in our lack of self-awareness. Our instincts are to survive and we will have the intelligence and imagination required to not be terminated by our own creations.

When Japanese robots replace kindergartners and raise children, when they are trusted to look after the most vulnerable in society by checking on babies' health in their cribs and improving the mental health of the lonely, abandoned elderly, where have we got to as societies?

We draw the line with the mental and emotional harming of our young people as seen with the deactivation of the pornographic Boyfriend Maker, yet we do not adequately protect youth online from random porn links, sexual approaches, and exposure to the most unsavory aspects of virtual life like mobbing and bullying, which clearly damages mental health to the point of people killing themselves.

I have been working on a prototype of a *Guardian Chatbot* to protect kids online and encourage them to go outdoors and play. Have a social life with real people. And they may well listen to this AI bot telling them do things they would refuse to do if told by their parents. The way to get into the head space of the next generation—for good not bad—is perhaps the all-pervasive Next Gen AI bot if it can be deployed for a higher purpose when concerning youth.

For adults, responsible for their own actions, Next Gen AI bots might be the way to influence behavior for the good of society, not ill. For this evolution to happen in the least controversial and most broadly acceptable way, in all varieties of multicultural contexts, we will need to be "on the

---

[26] For this reason, my Recommended Reading list at the end of this White Paper has only female authors, essayists, academics, industry commentators and tech bloggers in an attempt to create some gender balance.

[27] https://futureoflife.org/ai-open-letter/"Research Priorities for Robust and Beneficial Artificial Intelligence".

same page" for terms, conditions, regulations, and definitions.

Otherwise these emerging new lifeforms may just well take on a life of their own, independent of our human imagination and childlike fantasy. Ultimately, without adequately self-aware, carefully pragmatic measures in place, BaaS could develop commercially beyond our collective control and the individual chatbot's original, human-defined parameters for its online brain in the cloud.

Recommended Reading

Given the gross gender imbalance in the public debates and academic citations/conference invitations, I have decided to address this by giving you a woman-only reading list. This demonstrates that there is a lot of work and authoritative female voices who ought be included not excluded if indeed AI is to be controlled in every sphere so that it does not become rogue Super Intelligence and an existential threat to us all.

Women only writers and intellectuals on AI, chatbot programming and robots: Cultural Studies theorists and philosophers relevant to this emerging tech

Butler, Judith. American philosopher, feminist and gender theorist. She has been hugely influential internationally. Here are representative examples of her Cultural Studies theories.

Butler, J. 1990. *Gender Trouble: Feminism and the Subversion of Identity*. New York, NY: Routledge.

Butler, J. 1997. *Excitable Speech: A Politics of the Performative*. New York, NY: Routledge.

Butler, J. 2015. *Notes Toward a Performative Theory of Assembly*. Cambridge, MA: Harvard University Press.

Caliskan, A., J.J. Bryson, and A. Narayanan. 2017. "Semantics derived automatically from language corpora contain human-like biases". *Science* 356, no. 6334, pp. 183–186.

Chan, A. April 13, 2017. "Ai Picks Up Racial and Gender Biases When Learning from What Humans Write: There Is No Objectivity". *The Verge*. https://www.theverge.com/2017/4/13/15287678/machine-learning-language-processing-artificial-intelligence-race-gender-bias

Crowdflower.com 2017. Blog post. "The Gender of Artificial Intelligence" San Francisco. Machine Learning, chatbot and data mining venture. https://www.crowdflower.com/the-gender-of-ai/

Devlin, H. 13 April, 2017. "AI Programs Exhibit Racial and Gender Biases, Research Reveals". *The Guardian*. https://www.theguardian.com/technology/2017/apr/13/ai-programs-exhibit-racist-and-sexist-biases-research-reveals

Grebowicz, M., and H. Merrick. 2013. *Beyond the Cyborg: Adventures with Donna Haraway*. New York, NY: Columbia University Press

Grace, Katja. 2013. *Algorithmic Progress in Six Domains*. Berkeley, CA: Machine Intelligence Research Institute. https://intelligence.org/files/AlgorithmicProgress.pdf

Hayles, N.K. 1999. *How We Became Posthuman: Virtual Bodies in Cybernetics, Literature and Informatics*. Chicago, IL: University of Chicago Press.

Haraway, Donna. American professor of History of Consciousness and Feminist Studies at the University of California.

Haraway, D. 1991. *Symians, Cyborgs and Women: the Reinvention of Nature*. New York, NY: Routledge. Reprinted 2013. Includes her famous *Cyborg Manifesto: Science, Technology and Socialist-Feminism in the Late Twentieth Century* (1985).

Haraway, D. 2016. *Manifestly Haraway*. Minneapolis, MN: University of Minneapolis Press.

Kuchler, Hannah. A British finance and technology journalist. Correspondent in San Francisco for the Financial Times, London.

Lever, A.R., and D. Richards. 2017. "Robots in the family: Captain WH Richards and his mechanical men", pp. 70–85.

Lovelace, A. Biographies available online:
https://www.famousscientists.org/ada-lovelace/
https://findingada.com/about/who-was-ada/
https://iq.intel.com/ada-lovelace-the-first-computer-programmer/
http://www.bbc.co.uk/news/technology-34505896

Moosberg-Bustnes, H., and Loring, P. 2017. "Being human: Minds reflected in machines," pp. 18–31.

Morais, B. October 15, 2013. "Ada Lovelace, the First Tech Visionary". *The New Yorker*. https://www.newyorker.com/tech/elements/ada-lovelace-the-first-tech-visionary

Peitzker, T. Kindle ebooks. www.amazon.com/author/taniapeitzker

Peitzker, T. 2018. "Commercial Negotiations: Case Studies & Role Plays". (forthcoming).

Peitzker, T. 2019. *Uses and Risks of Chatbots in Business: Botification Within Applied Artificial Intelligence and Machine Learning*. New York, NY: BEP. (forthcoming).

Lecture on AI and bots, 2017, University of Cambridge Judge Business School. This audio recording of my talk at CBJS includes the SWAY presentation link https://youtu.be/f2seNpAnIEs

Public Lecture on chatbot evolution, 2016, University of Kent, Canterbury Campus. https://player.kent.ac.uk/Panopto/Pages/Viewer.aspx?id=186a5d61-ee1a-4bc3-9700-13a74dd1698f

Riskin, J., ed. 2007. *Genesis Redux: Essays in the History and Philosophy of Artificial Life*. Chicago, IL: Uni of Chicago Press.

Rossini, Manuela. A Swiss feminist professor of Anglistik, Spanish Philology Cultural Studies, about to publish a collection titled "A Genealogy of Posthumanisms" (2018). She is also the President of the SLSAeu, European Society for Literature Science and the Arts. www.slsa-eu.org

Rossini, M, and B. Clarke, eds. 2017. *The Cambridge Companion to Literature and the Posthuman*. Cambridge, UK: Cambridge University Press.

Russell, B, ed. 2017. *Robots: The 500 Year Quest to Make Machines Human. Science Museum*. London, UK: Scala Arts & Heritage Publishers. Exhibition accompanying book. Includes these essays by women experts.

Singh, H. 9 July, 2017. "Can Artificial Intelligence Usher an Era of Gender Parity". Filmmaker and guest writer Huffpost.

Suchman, L. 2007. *Human-Machine Reconfigurations*. New York, NY: Cambridge University Press.

Toffoletti, K. 2007. *Cyborgs and Barbie Dolls: Feminism, Popular Culture and the Posthuman Body*. London, UK: IB Tauris Publisher.

Truit, E.R. 2015. *Medieval Robots: Mechanism, Magic, Nature and Art*. Philadelphia, PA: Uni of Pennsylvania Press.

Truitt, E.R. 2017. "In whose image? Ancient and medieval automata," pp. 32–47.

Vaid, Prachi. Forthcoming in 2018, Masters research paper. University of Cambridge Judge Business School. Title TBC. This Masters student attended my talk on AI and bots in October 2017. I was subsequently interviewed by Prachi for her research thesis.

Voskuhl, A. 2013. *Androids in the Enlightenment: Mechanics, Artisans and Cultures of the Self*. Chicago, IL: Uni Chicago Press.

Wajcman, J. 2015. *Pressed for Time: The Acceleration of Life in Digital Capitalism*. Chicago, IL: University of Chicago Press.

Wajcman, J. 2017. "Humanoid robots and the promise of an easier life," pp. 102–117.

www.ingramcontent.com/pod-product-compliance
Lightning Source LLC
Chambersburg PA
CBHW040352220526
45473CB00009B/2860